STONE & STEEL

A · LOOK · AT · ENGINEERING · BY · GUY · BILLOUT

PRENTICE-HALL, INC. • ENGLEWOOD CLIFFS, NEW JERSEY

Printed in the United States of America.

Prentice-Hall International, Inc., London
Prentice-Hall of Australia, Pty. Ltd., North Sydney
Prentice-Hall of Canada, Ltd., Toronto
Prentice-Hall of India Private Ltd., New Delhi
Prentice-Hall of Japan, Inc., Tokyo
Prentice-Hall of Southeast Asia Pte. Ltd., Singapore
Whitehall Books Limited, Wellington, New Zealand

10 9 8 7 6 5 4 3 2 1

Library of Congress Cataloging in Publication Data

Billout, Guy.
Stone and Steel.

SUMMARY: Text and illustrations describe bridges and buildings of historic interest.
1. Civil engineering—Juvenile literature.
2. Bridges—Juvenile literature. 3. Buildings—Juvenile literature. [1. Civil engineering.
2. Bridges. 3. Buildings.] I. Title.
TA149.B54 1980
ISBN 0-13-846873-7

STONE & STEEL

This book is dedicated to the children I love
(in alphabetical order)
Daniel, Elisa, Frédéric, Gaby, Gaël, Grégoire,
Guillaume, Julia, Juliette, Melissa, Rachel P., Rachel W.,
Romain G., Romain L., Rodolphe, Samuel, Serge,
Sonia, Stéphanie, Vincent, Yves.

THE CORINTH CANAL IN GREECE

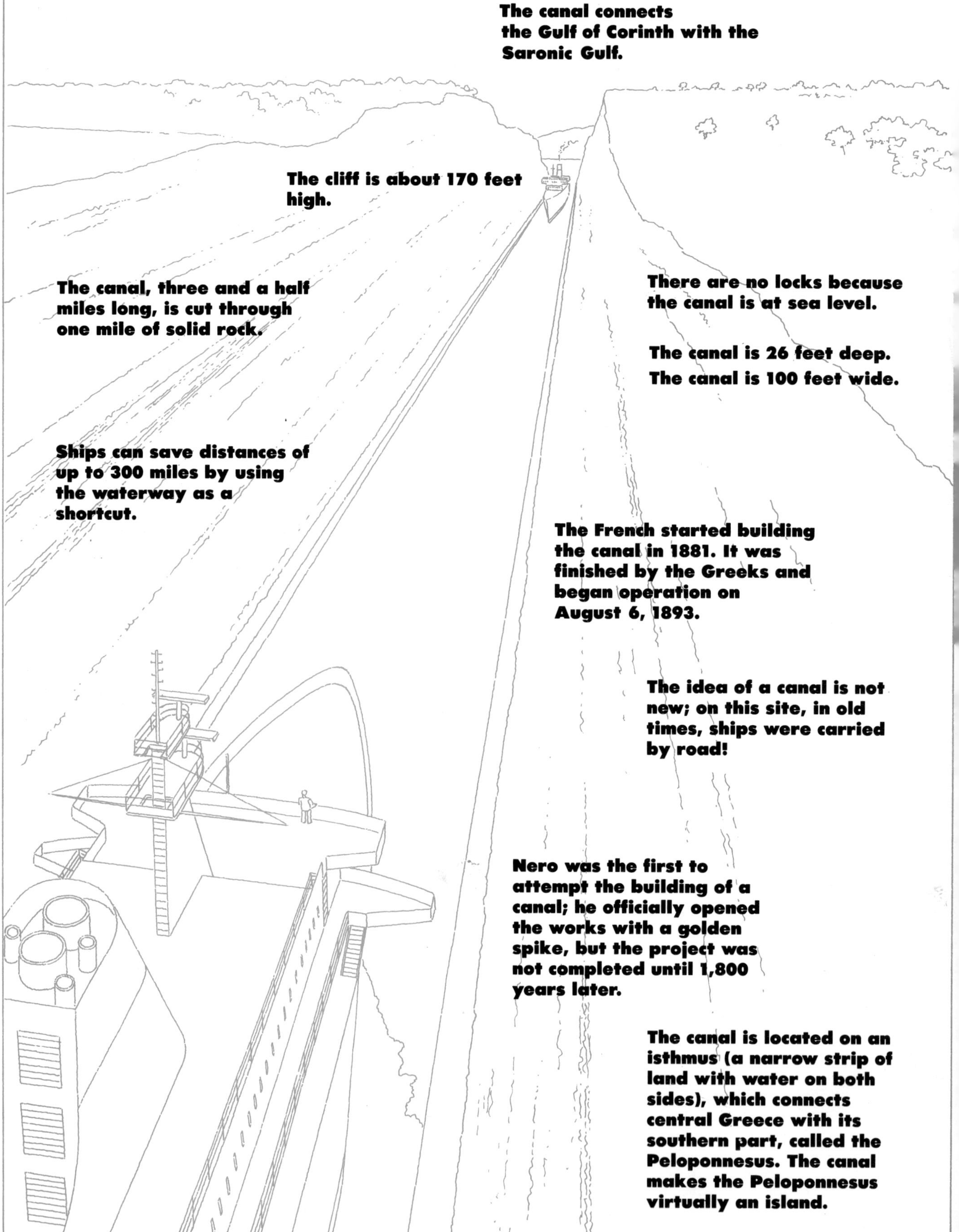

The canal connects the Gulf of Corinth with the Saronic Gulf.

The cliff is about 170 feet high.

The canal, three and a half miles long, is cut through one mile of solid rock.

There are no locks because the canal is at sea level.

The canal is 26 feet deep.

The canal is 100 feet wide.

Ships can save distances of up to 300 miles by using the waterway as a shortcut.

The French started building the canal in 1881. It was finished by the Greeks and began operation on August 6, 1893.

The idea of a canal is not new; on this site, in old times, ships were carried by road!

Nero was the first to attempt the building of a canal; he officially opened the works with a golden spike, but the project was not completed until 1,800 years later.

The canal is located on an isthmus (a narrow strip of land with water on both sides), which connects central Greece with its southern part, called the Peloponnesus. The canal makes the Peloponnesus virtually an island.

THE MINOTS LEDGE LIGHTHOUSE, OFF THE BOSTON BAY, U.S.A.

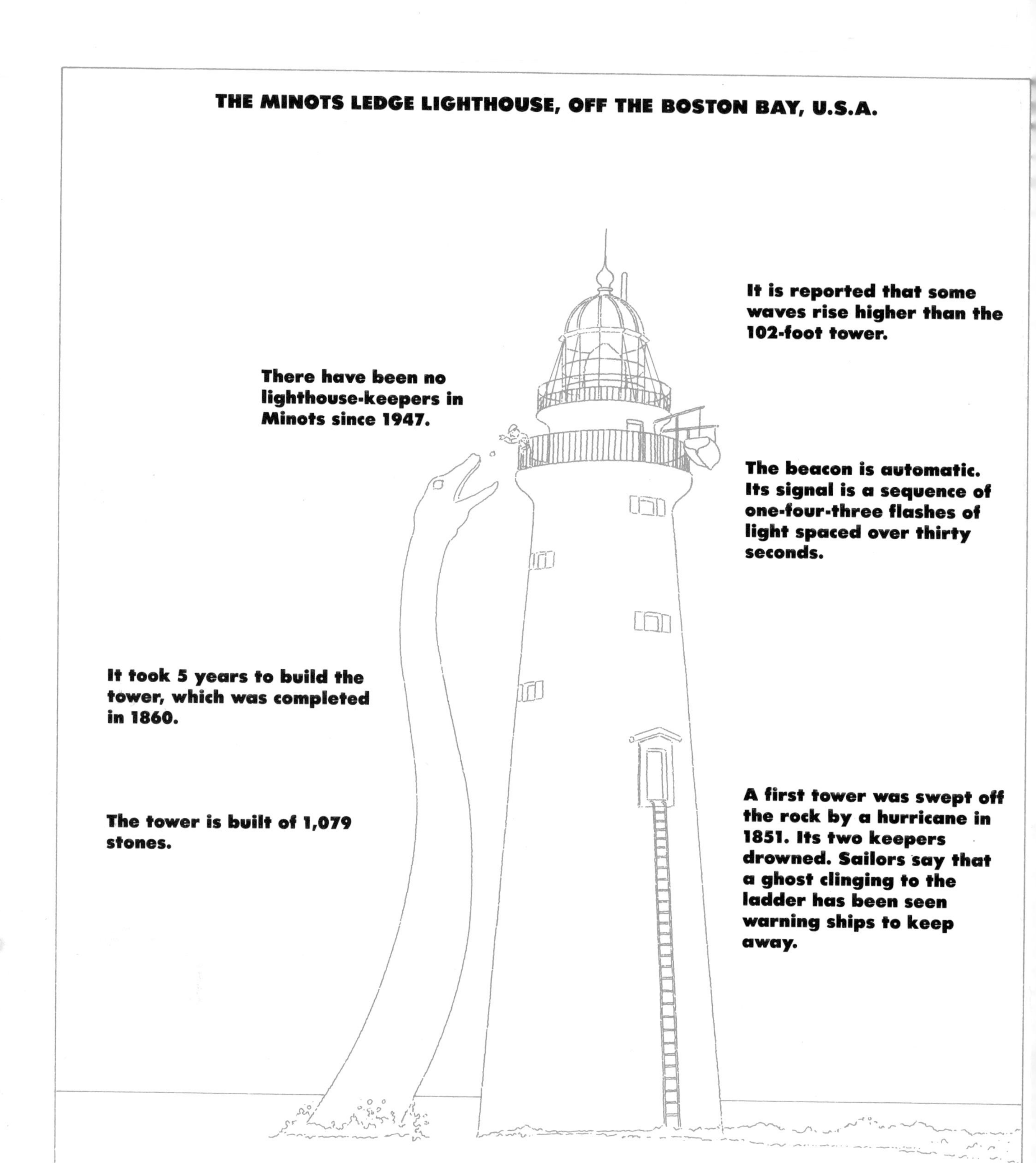

The tower rests on rocks which are submerged most of the time.

Between 1832 and 1841 there were forty ship-wrecks on Minots Ledge and the nearby rocks.

THE SAMRAT YANTRA AT THE JAIPUR OBSERVATORY, IN INDIA

Samrat Yantra, "the supreme instrument," was built to allow verification of the calculations and observations of astronomers of the time.

This observatory was built in 1724.

Telescopes, invented 100 years before the building of the observatory, were not used in India. Small brass instruments which gave imprecise readings were used. It was hoped that making the instruments larger would increase the accuracy of the readings, but this was not the case.

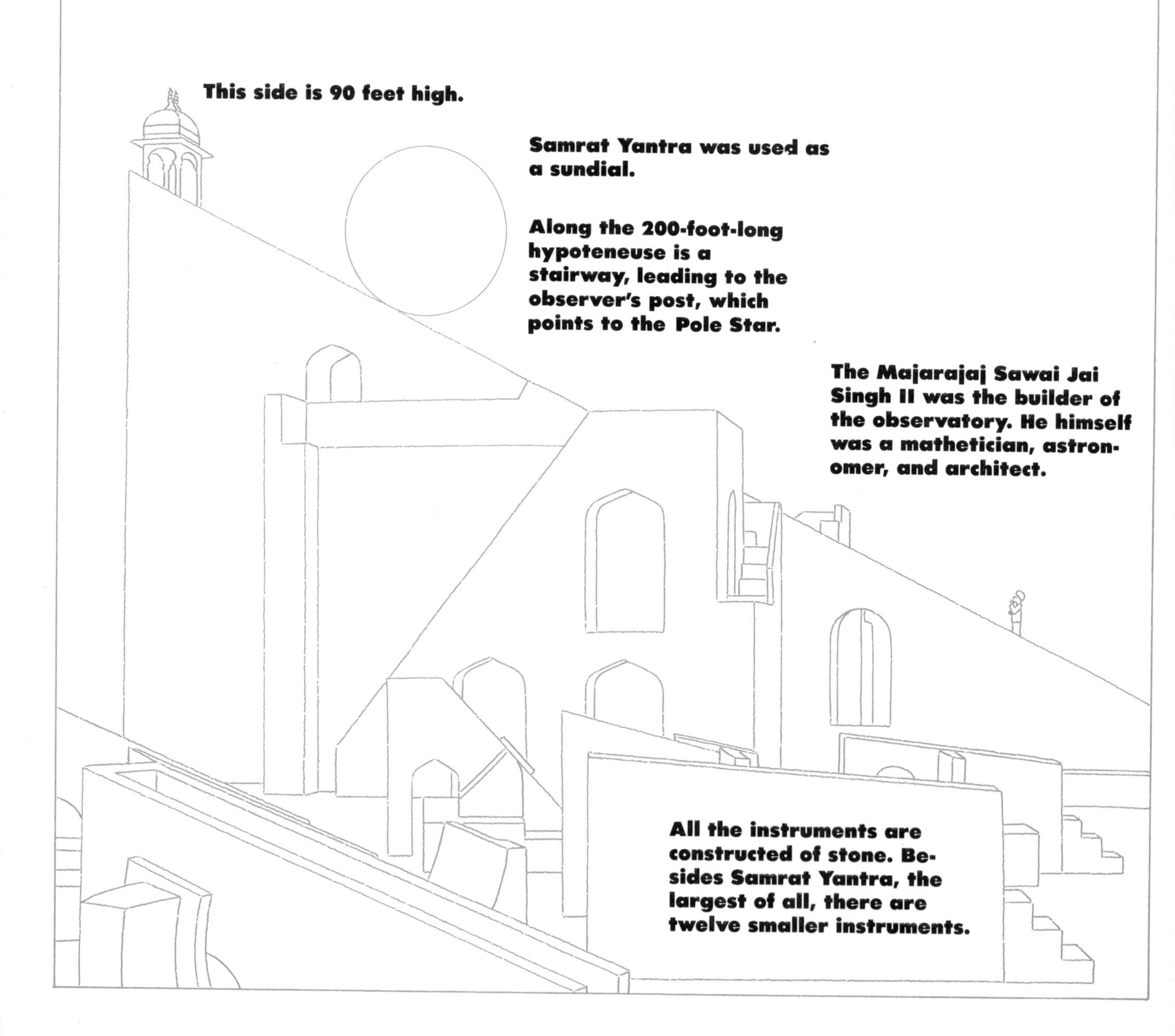

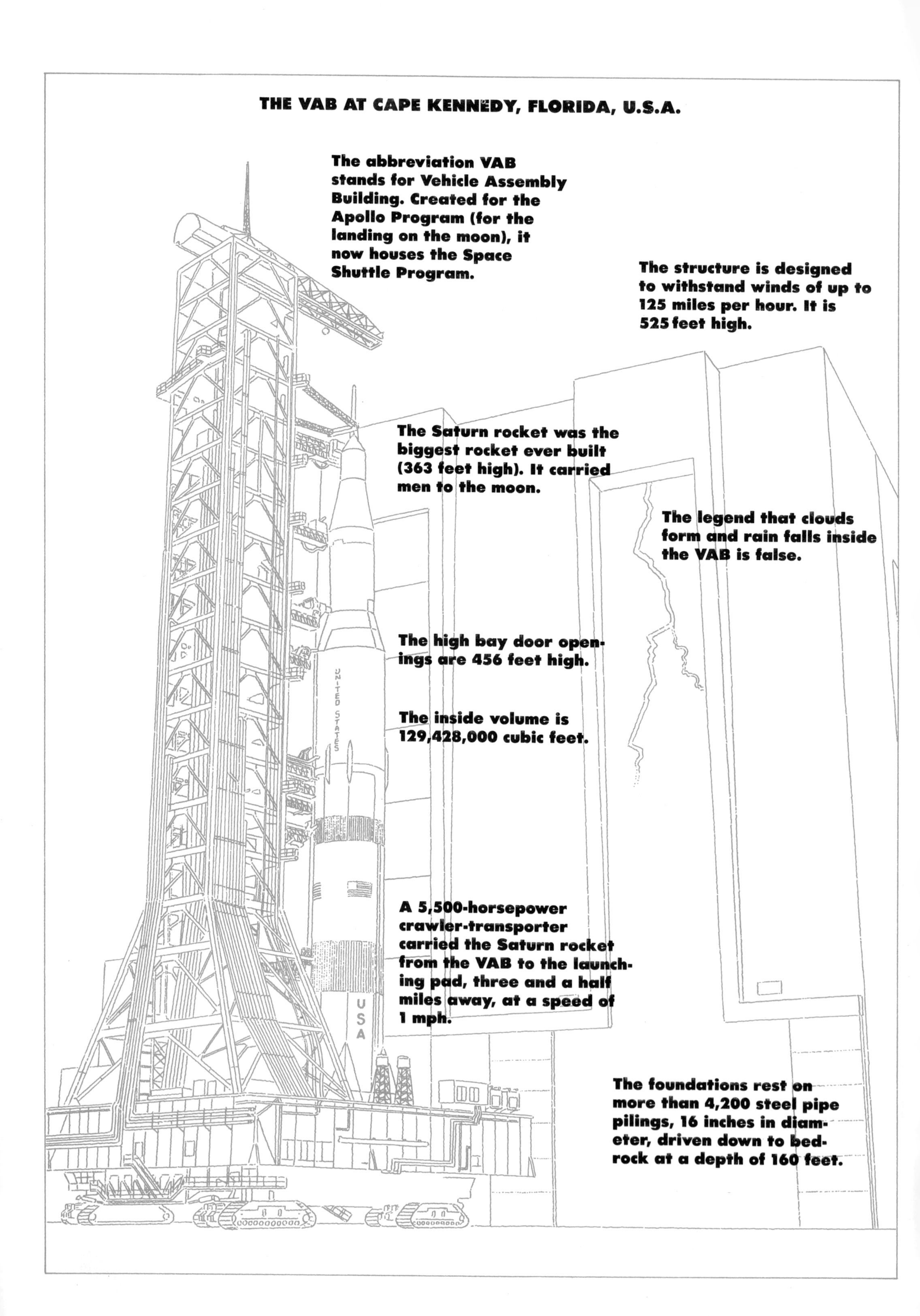
THE VAB AT CAPE KENNEDY, FLORIDA, U.S.A.
The abbreviation VAB stands for Vehicle Assembly Building. Created for the Apollo Program (for the landing on the moon), it now houses the Space Shuttle Program.
The structure is designed to withstand winds of up to 125 miles per hour. It is 525 feet high.
The Saturn rocket was the biggest rocket ever built (363 feet high). It carried men to the moon.
The legend that clouds form and rain falls inside the VAB is false.
The high bay door openings are 456 feet high.
The inside volume is 129,428,000 cubic feet.
A 5,500-horsepower crawler-transporter carried the Saturn rocket from the VAB to the launching pad, three and a half miles away, at a speed of 1 mph.
The foundations rest on more than 4,200 steel pipe pilings, 16 inches in diameter, driven down to bedrock at a depth of 160 feet.
UNITED STATES
USA

UNITED STATES
USA

THE CHESAPEAKE BAY BRIDGE-TUNNEL IN VIRGINIA, U.S.A.

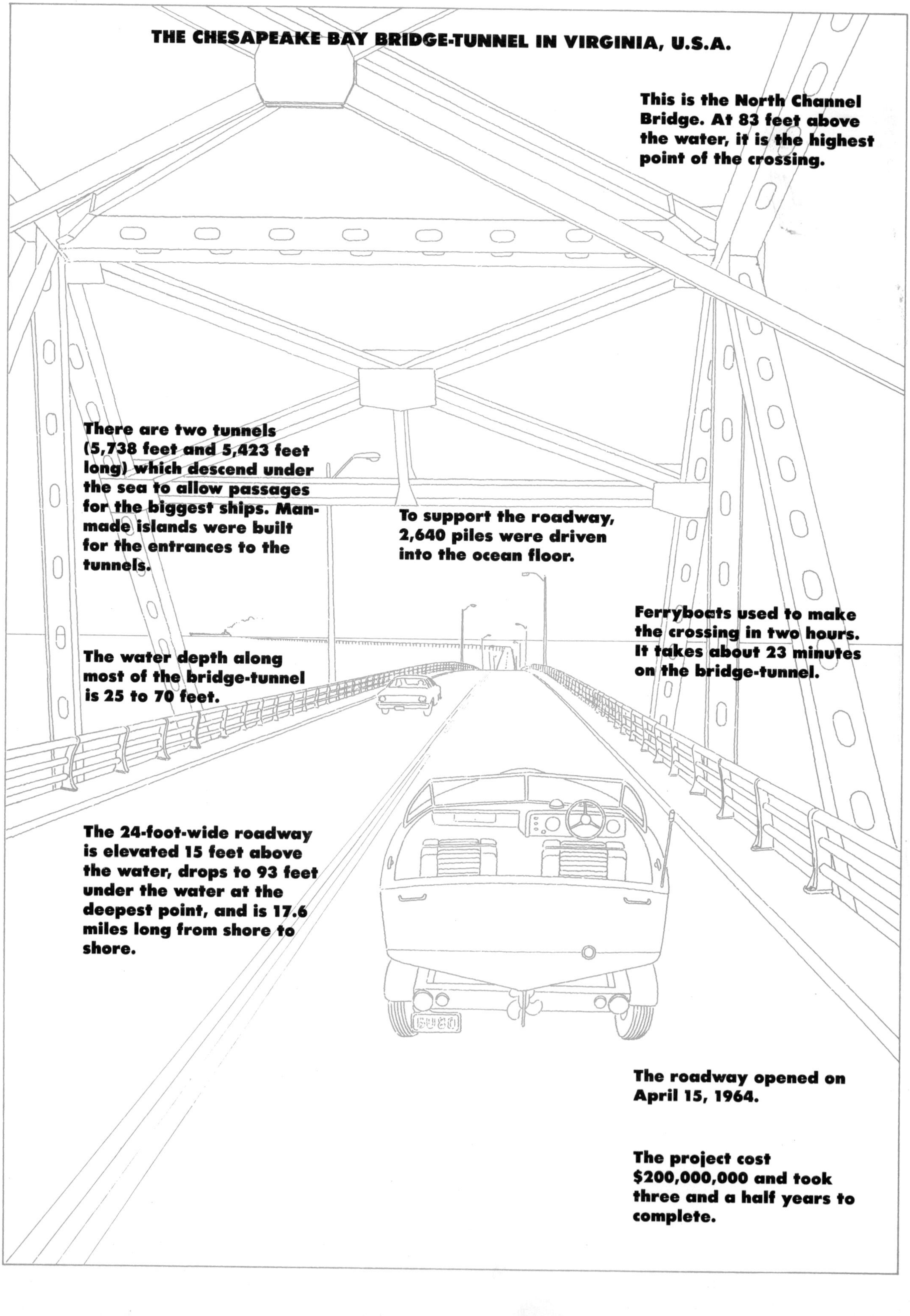

This is the North Channel Bridge. At 83 feet above the water, it is the highest point of the crossing.

There are two tunnels (5,738 feet and 5,423 feet long) which descend under the sea to allow passages for the biggest ships. Man-made islands were built for the entrances to the tunnels.

To support the roadway, 2,640 piles were driven into the ocean floor.

Ferryboats used to make the crossing in two hours. It takes about 23 minutes on the bridge-tunnel.

The water depth along most of the bridge-tunnel is 25 to 70 feet.

The 24-foot-wide roadway is elevated 15 feet above the water, drops to 93 feet under the water at the deepest point, and is 17.6 miles long from shore to shore.

The roadway opened on April 15, 1964.

The project cost $200,000,000 and took three and a half years to complete.

THE PLALZGRAFENSTEIN CASTLE ON THE RHINE, GERMANY

THE DAKHMAS OR TOWERS OF SILENCE IN INDIA

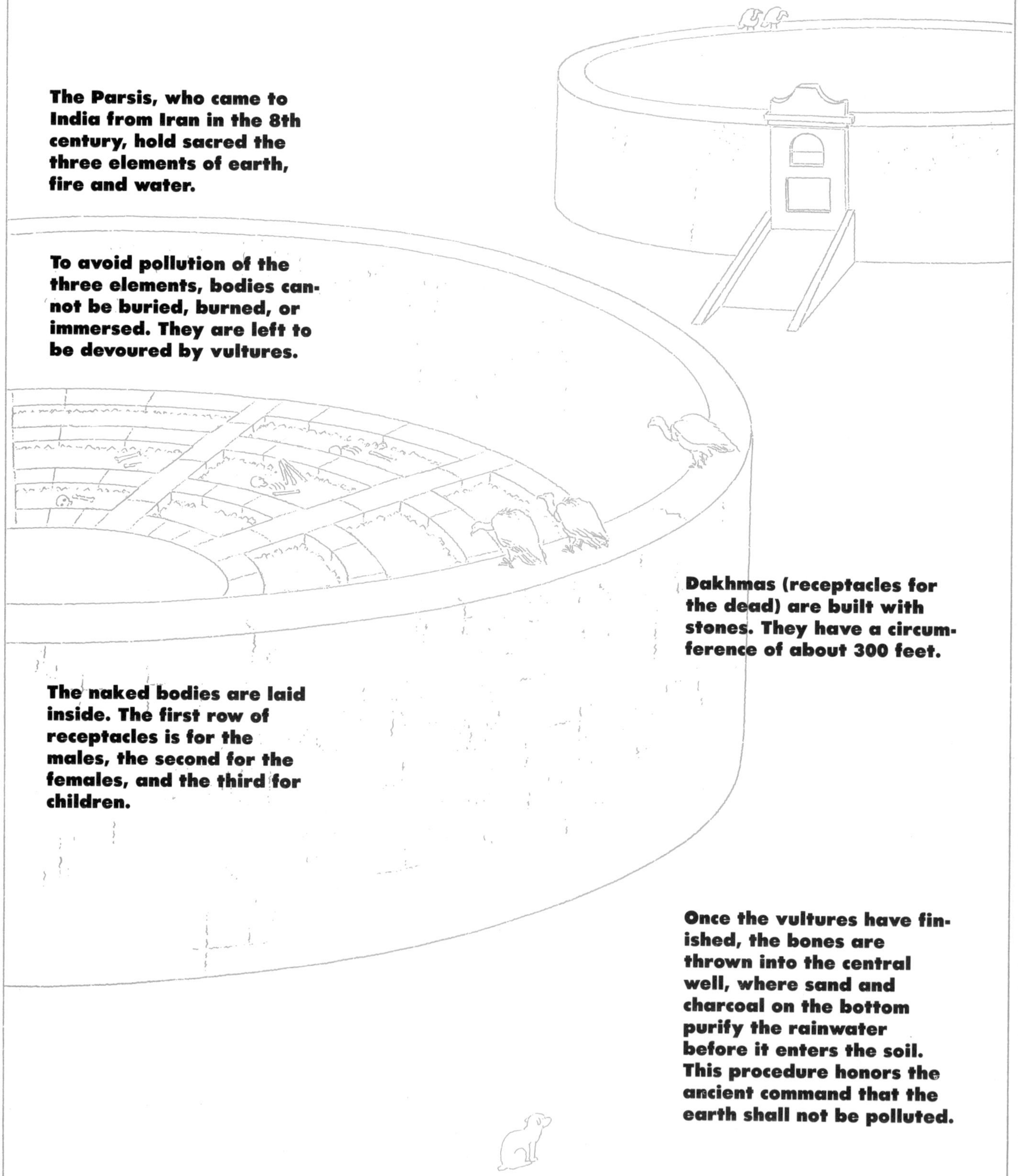

The Parsis, who came to India from Iran in the 8th century, hold sacred the three elements of earth, fire and water.

To avoid pollution of the three elements, bodies cannot be buried, burned, or immersed. They are left to be devoured by vultures.

Dakhmas (receptacles for the dead) are built with stones. They have a circumference of about 300 feet.

The naked bodies are laid inside. The first row of receptacles is for the males, the second for the females, and the third for children.

Once the vultures have finished, the bones are thrown into the central well, where sand and charcoal on the bottom purify the rainwater before it enters the soil. This procedure honors the ancient command that the earth shall not be polluted.

Dogs are sacred animals for the Parsis. A deceased cannot enter the dakhma without being looked at by a dog in a ceremony called "segdid."

THE PANAMA CANAL LOCKS

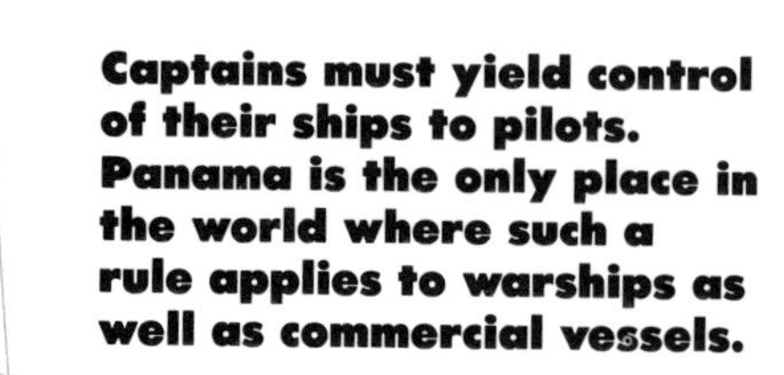

Captains must yield control of their ships to pilots. Panama is the only place in the world where such a rule applies to warships as well as commercial vessels.

Electric locomotives called mules help move a ship through a lock. Large vessels need six of them.

On both ends of the canal locks raise ships to 85 feet above sea level and, of course, lower them as well.

For each passage of a ship through the canal some 52 million gallons of water spill out of the locks.

The water supply for the locks comes from a lake, which in turn is supplied by tropical rains.

In 1881 the French started digging the canal and abandoned it 9 years later. Twenty thousand workers died, mostly of malaria and yellow fever.

The Americans resumed the digging of the canal in 1904 and completed it in 1914.

Five percent of the world's total ocean cargo passes through the Panama Canal.

Non-commercial ships are charged about a dollar per ton.

More and more warships and supertankers are too large for the canal. Plans for a sea level canal are impaired by some ecolog-ical problems and by a $5.3 billion estimated cost.

US
7

THE BOONE WINDMILL IN NORTH CAROLINA, U.S.A.

The tower is 140 feet high.

The propeller-type blades span 200 feet, more than a Boeing 747.

The rotor turns at 34 revolutions per minute (rpm) and drives, through a transmission train, an 1800-revolution per minute alternator.

The turbine shuts down in winds under 11 miles per hour and above 35 miles per hour, but it is designed to withstand winds of 150 miles per hour. In winds between 25 and 35 miles per hour, the wind turbine will produce 2,000 kilowatts —enough power for 500 homes.

DOE NASA

At the time of its completion, this mill was the largest in the world. Machines more advanced and even larger are currently being designed and developed.

The entire structure weighs 650,000 pounds.

DOE
NASA

THE CATHEDRAL OF LEARNING IN PITTSBURGH, PENNSYLVANIA

The tower is 52 stories high—680 feet tall, and has 3,000 windows.

The building uses 6,000,000 kilowatts of electricity per year.

Built with limestone from Indiana, the cathedral was begun in 1926 and finished in 1937—at a cost of $34,000,000.

Charles Z. Klauder is the architect.

Inside are classrooms, libraries, shops, laboratories and recreation centers. With the present permanent buildings of the university, it can accommodate 12,000 students and 1,800 employees.

The building at its base is 260 feet long, 260 feet wide. The foundation rests on bedrock 150 feet below.

THE BATTERSEA POWER STATION IN LONDON, ENGLAND

The main buildings are constructed of brick. They were built between 1929 and 1935.

Sir Giles Gilbert Scott was the architect.

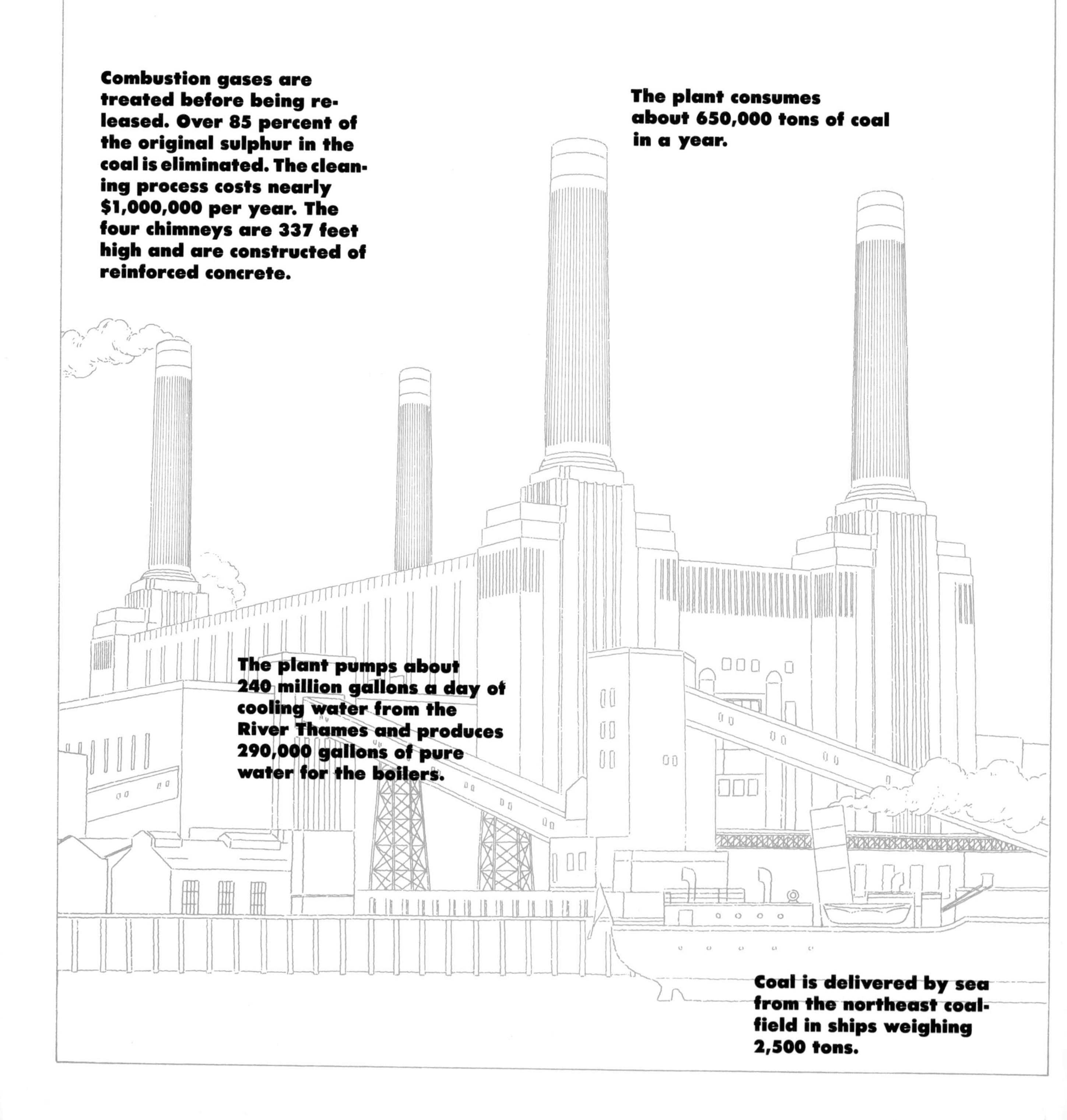

Combustion gases are treated before being released. Over 85 percent of the original sulphur in the coal is eliminated. The cleaning process costs nearly $1,000,000 per year. The four chimneys are 337 feet high and are constructed of reinforced concrete.

The plant consumes about 650,000 tons of coal in a year.

The plant pumps about 240 million gallons a day of cooling water from the River Thames and produces 290,000 gallons of pure water for the boilers.

Coal is delivered by sea from the northeast coalfield in ships weighing 2,500 tons.

THE HOOVER DAM ON THE COLORADO RIVER, U.S.A.

Begun in 1931, the dam was completed in 1935, two years ahead of schedule.

The towers are 395 feet high.

The roadway is over 700 feet above the foundation rock.

There is enough concrete in the dam to pave a 16-foot-wide highway between San Francisco and New York.

In 1941, Lake Mead reached its normal level—as it is today. The lake is 580 feet deep, and 115 miles long.

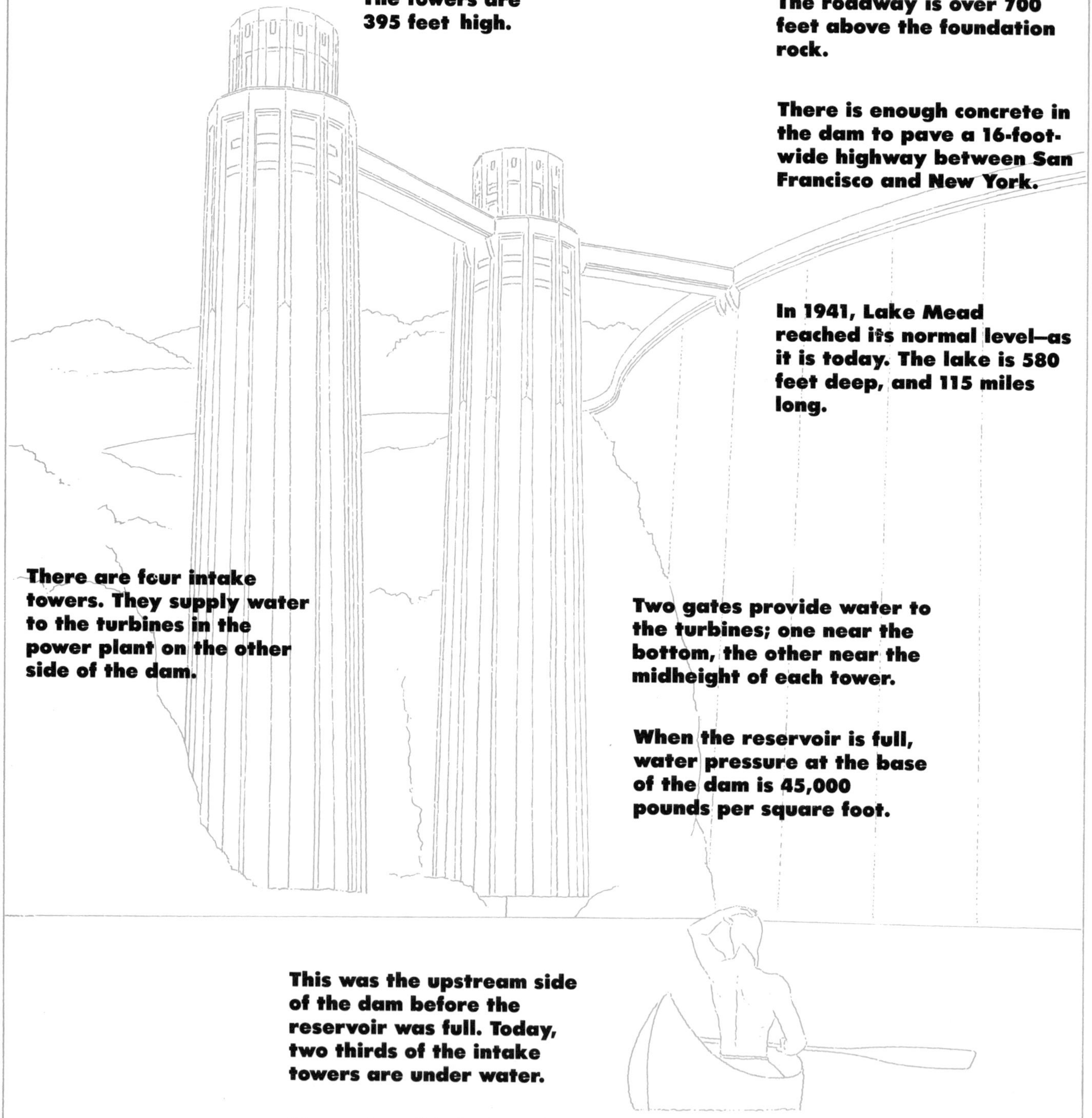

There are four intake towers. They supply water to the turbines in the power plant on the other side of the dam.

Two gates provide water to the turbines; one near the bottom, the other near the midheight of each tower.

When the reservoir is full, water pressure at the base of the dam is 45,000 pounds per square foot.

This was the upstream side of the dam before the reservoir was full. Today, two thirds of the intake towers are under water.

The water of Lake Mead, through Hoover Dam's turbines, provides 1,344,800 kilowatts, or 1,857,000 horsepower.

THE MICHIGAN AVENUE BRIDGE IN CHICAGO, ILLINOIS, U.S.A.

The Michigan Avenue Bridge is a bascule bridge. "Bascule" is the French word for "teeter totter" and refers to the principle of balancing the bridge by using a system of counterweights.

The bridge cost $14 million when it was built in 1920.

Traffic regulations require that the bridge stay down at least 10 minutes between lifts.

There is an upper roadway for passenger vehicles and pedestrians, and a lower roadway for heavy commercial trucking and pedestrians.

This is the Wrigley Building built with a chewing gum fortune.

It takes one minute to open the bridge.

It takes about six minutes for a vessel to proceed.

There are 60 bridges on the Chicago River; 56 of them are bascule bridges.

Although each half of the bridge weighs 7½ million pounds, it is so delicately balanced that a 100-horsepower motor is enough to lift it.

608

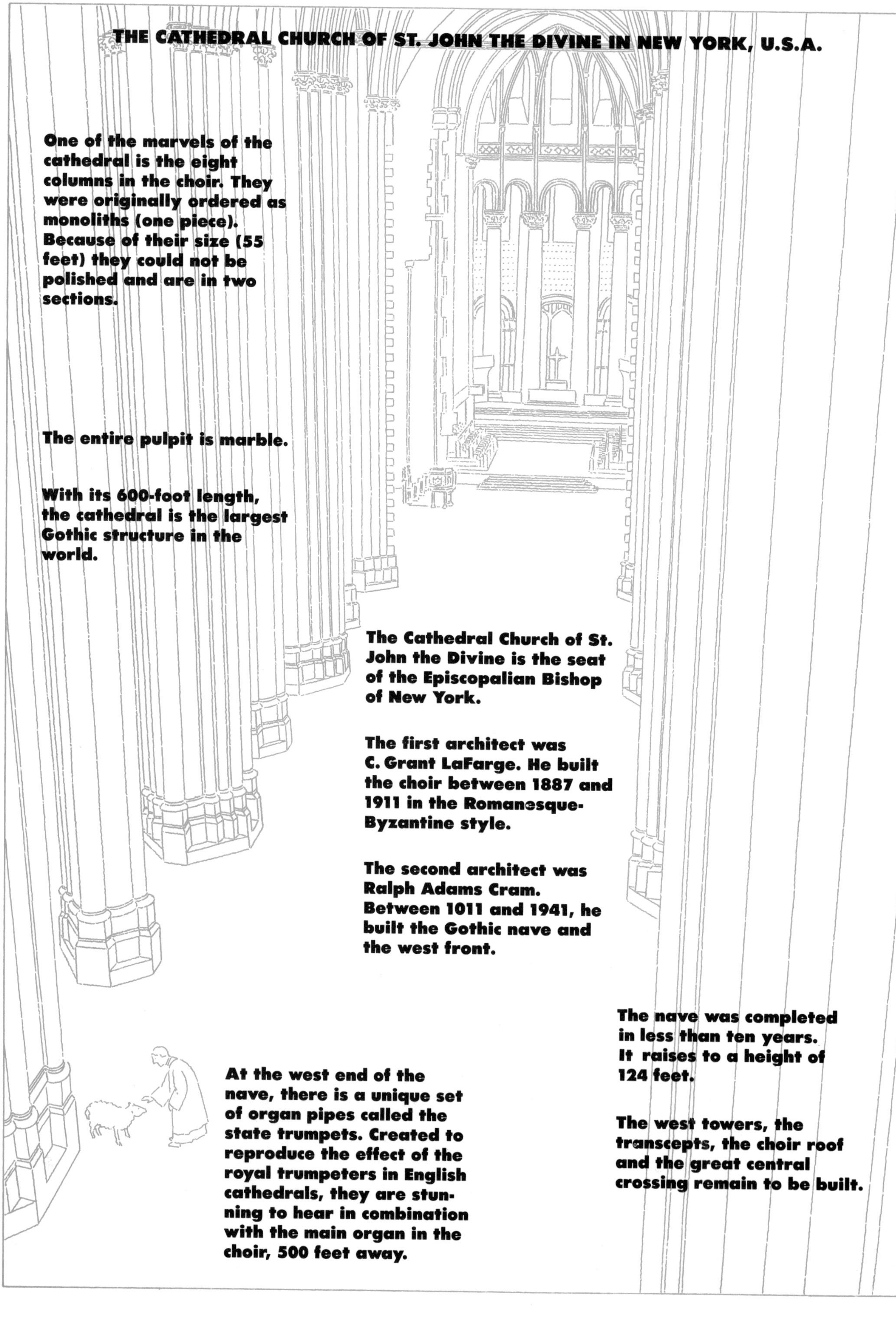

THE CATHEDRAL CHURCH OF ST. JOHN THE DIVINE IN NEW YORK, U.S.A.

One of the marvels of the cathedral is the eight columns in the choir. They were originally ordered as monoliths (one piece). Because of their size (55 feet) they could not be polished and are in two sections.

The entire pulpit is marble.

With its 600-foot length, the cathedral is the largest Gothic structure in the world.

The Cathedral Church of St. John the Divine is the seat of the Episcopalian Bishop of New York.

The first architect was C. Grant LaFarge. He built the choir between 1887 and 1911 in the Romanesque-Byzantine style.

The second architect was Ralph Adams Cram. Between 1011 and 1941, he built the Gothic nave and the west front.

The nave was completed in less than ten years. It raises to a height of 124 feet.

The west towers, the transcepts, the choir roof and the great central crossing remain to be built.

At the west end of the nave, there is a unique set of organ pipes called the state trumpets. Created to reproduce the effect of the royal trumpeters in English cathedrals, they are stunning to hear in combination with the main organ in the choir, 500 feet away.

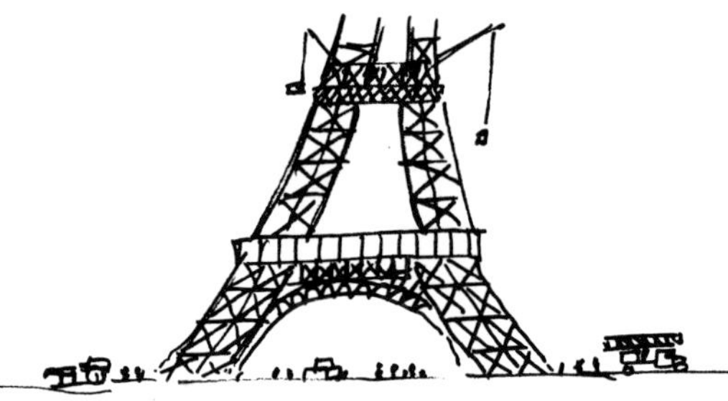

I gathered many photographs of the different subjects of this book. Laying a transparent film on each photograph, I copied what I felt to be the most interesting part of the picture. Transferring the resulting drawing onto tracing paper, I recorded it on the final board to ink it. At that point, it looked like the illustrations as they appear on the left hand pages of the book. Then, using Winsor and Newton water colors with an Isabey #2 sable brush and a Paasche air-brush, I colored the drawings. What I found interesting about this technique was the transformation of the photograph into a drawing: some photographs first appeared to me too banal to make good pictures, but as soon as I made line drawings of them, they became something of their own. The addition of color dramatized the transformation even more.

Even though the technical details are as exact as I could make them, this book is not a reference book; my main purpose was to convey a poetic mood.